© **Copyright 2021 Small Footprint Press - All rights reserved**.

The content contained within this book may not be reproduced, duplicated or transmitted without direct written permission from the author or the publisher.

Under no circumstances will any blame or legal responsibility be held against the publisher, or author, for any damages, reparation, or monetary loss due to the information contained within this book, either directly or indirectly.

Legal Notice:

This book is copyright protected. It is only for personal use. You cannot amend, distribute, sell, use, quote or paraphrase any part, or the content within this book, without the consent of the author or publisher.

Disclaimer Notice:

Please note the information contained within this document is for educational and entertainment purposes only. All effort has been executed to present accurate, up to date, reliable, complete information. No warranties of any kind are declared or implied. Readers acknowledge that the author is not engaged in the rendering of legal, financial, medical or professional advice. The content within this book has been derived from various sources. Please consult a licensed professional before attempting any techniques outlined in this book.

By reading this document, the reader agrees that under no circumstances is the author responsible for any losses, direct or indirect, that are incurred as a result of the use of the information contained within this document, including, but not limited to, errors, omissions, or inaccuracies.

———————————

Introduction

When we talk about family safety and home security, top-of-mind concerns typically involve household or crime accidents. Nonetheless, environmental factors force millions of individuals to evacuate their shelters or homes every year.

In 2020, there were a total of 416 natural disaster events worldwide. The Asian Pacific region suffered the second-highest number of natural disasters, most likely because of its susceptibility and size.

Emergency preparedness is not about a group of survivalists swarming around in the forests, preparing to fight off the hungry hordes in some apocalyptic fantasy. In today's world of superstorms, global warming, financial crises and terrorist attacks, we acknowledge that most people will probably experience significant changes in the flow of goods and electricity at some point in our lives.

Keeping up some extra supplies, understanding some new skills, and creating emergency contingency plans does not take a considerable amount of money and time. Furthermore, it is a low-cost insurance that could foster peace of mind, especially in turbulent times.

No one knows what the future will bring to us. You cannot plan for all possible situations. However, an intelligent person plans for several of the most likely possibilities and keeps at least some basic supplies for emergencies.

Indeed, emergency workers will help after a disaster strike. However, they may not be able to reach everybody immediately. That's why it is essential to be ready to survive on your own for at least three days during an emergency.

That may suggest having another place to stay, extra water, food, first-aid, and other basic needs. We know the fact that we cannot control emergencies, natural disasters, or terrorist attacks. But we can be ready for them!

This is where this guide comes in. It is intended to help you prepare by learning how to plan for emergencies.

---

1 Jaganmohan M. (2021). Annual number of natural disaster events globally from 2000 to 2020. Statista. Retrieved from https://www.statista.com/statistics/510959/number-of-natural-disasters-events-globally/

Remember that the first thing you can do to safeguard yourself, your family, and your home from a natural disaster is to plan how you will handle the disaster well before it approaches. With a plan in place, you can act immediately when disaster strikes. Read this guide, and you will learn how to:

- Protect your loved ones from harm
- Safeguard your belongings
- Create an emergency kit
- Make an emergency plan
- Prepare your home to weather the storm
- Help those affected by natural disasters
- Prepare your finances

As an advocate of sustainable living, Small Footprint Press aims to spread awareness and allow people—like you—to make small and big changes in your life that will help you live more sustainably and self-sufficient. We strive to help people accomplish a sustainable lifestyle, enabling them to give back to Mother Earth.

Hence, we dedicated this book to you. We will cover different subjects on what you need to do before a disaster to guarantee you are protected. Even if you don't have any idea about emergency preparedness, this book will equip you with all the information you need to get started.

Chapter 1:
GET STARTED

Let's face it. Getting started on emergency preparedness could sometimes feel vast and overwhelming. But it doesn't have to be. We believe that you'll become more successful in the long run by continuously doing small things.

Life is busy and hectic. There never seems to be enough time in the day to get the things you need to get done, let alone prepare for a disaster. However, you will find small, simple things you can do to move closer to being prepared for emergencies or big disasters.

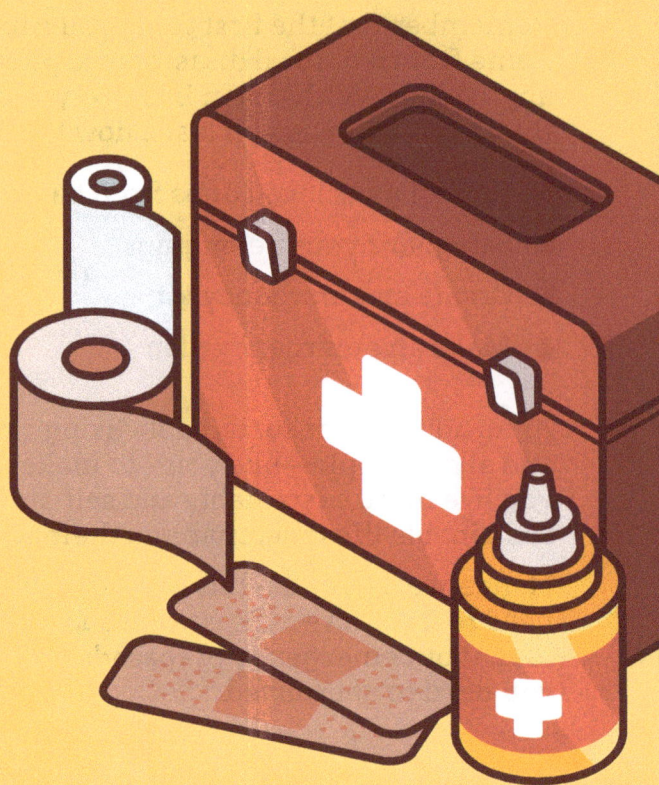

Get Your Kit Together

Prepare an emergency kit that includes all the medication, water, food, and supplies you'll need to cope with for at least seven days without outside help. You can buy ready-made kits at your local store, or you can make your kit over time.

Make an Emergency Plan

Chances are you and your family won't be together when a major disaster happens (like an earthquake). Having an emergency plan and strategies to reconnect with loved ones and family is crucial after a disaster. Setting up a meeting location and an out-of-area contact person are two critical ways to reconnect.

It also pays to have contacts written down when the power goes out, and there's nowhere to charge your mobile phone. Create a small contact list with numbers that everybody could keep in their wallets. You can also leave a copy in your emergency kit.

In case local lines get jammed, create a plan for checking in with relatives. Text messages will sometimes go through, even if phone lines are clogged.

The Prepping Mentality

One of the critical things you need to do is to accept that a catastrophe could strike anytime. It indicates accepting responsibility for your security and safety and that of your family. It implies being active in the face of an emergency instead of being passive.

Take some time to find out and research the most likely disasters that could happen. These include economic collapse, earthquakes, flooding, hurricanes, wildfires, storms, and EMP attacks.

Of course, some of these threats are highly unlikely. However, "highly unlikely" does not suggest it won't occur. Accept that there's a real risk, and you'll be in a better position psychologically when a catastrophe does strike.

Stockpile Emergency Supplies

Keep enough supplies in your home or in your car to meet your needs for approximately three days—a 3-day supply of water (one gallon per person every day) and food that won't spoil. Don't forget the non-food items too, like toothpaste, toilet paper, etc.

As with any stockpile of food, you can buy, then eat, then replenish and start over. After all, you do not like a stockpile of emergency food that's well beyond its expiration date.

One of the most vital parts of storing food is ensuring you use it and replace it often, so you don't end up with a pantry loaded with wasted food.

Chapter 2:
BE FINANCIALLY PREPARED

Financial preparedness is an integral part of emergency readiness. When disaster strikes, an emergency savings fund could be one of the major resources to deal with natural disasters and jumpstart recovery.

Recent studies have proven that Americans are underprepared for emergencies and the sudden costs, which come with them. In fact, approximately fifty-five percent of Americans think about an unplanned financial emergency.

This section will discuss specific actions that you and your families can take to become financially ready for any emergency.

Northwestern Mutual. (2018). New Research: Money Is The Leading Source Of Happiness --- And Stress. Northwestern Mutual. Retrieved from https://news.northwesternmutual.com/2018-06-12-New-Research-Money-Is-The-Leading-Source-Of-Happiness-And-Stress

Create an Emergency Savings Fund

Layoffs and furloughs are typical during disasters as organizations seek ways to stay afloat and cut costs. It would help if you had a cushion that could supplement your salary—or replace it even in the worst-case scenario.

You can strive to have six months of salary to continue to cover necessary expenses while you look for a new job. Setting aside three months of income is still a good start.

Gather Copies of Important Documents

Keep enough supplies in your home or in your car to meet your needs for approximately three days—a 3-day supply of water (one gallon per person every day) and food that won't spoil. Don't forget the non-food items too, like toothpaste, toilet paper, etc.

As with any stockpile of food, you can buy, then eat, then replenish and start over. After all, you do not like a stockpile of emergency food that's well beyond its expiration date.

One of the most vital parts of storing food is ensuring you use it and replace it often, so you don't end up with a pantry loaded with wasted food.

As we live in a modern and digital age, it's all too simple to keep your important financial documents online. Back up receipts, insurance information, and tax returns through a cloud storage system so you can access them from any device. Ensure you have physical copies as well.

You will need these physical copies kept safe in a waterproof or fireproof container in the event of a disaster that causes you to lose power or internet connection.

Here are some documents you need to find and print out:

- Power of attorney papers
- Social Security cards
- Insurance policies
- Deeds and ownership forms
- Medical information
- Marriage certifications, child support documents, and prenuptial agreements
- A living will and last will testament
- Passports, IDs, adoption papers and birth certificates,

Document Your Important Valuables

Make a detailed home inventory that consists of everything inside your home, from clothing to kitchen appliances and furniture. Even though you don't have any expensive items, the cumulative value of your things could add up fast.

Home insurance is created to safeguard not just your home's physical structure but also the items inside. Having documentation on your belongings will make the process go more smoothly if you ever need to file a claim.

Chapter 3:
STORE WATER SUPPLY

One of the priorities for emergency preparedness is water storage. Clean and safe drinking water is necessary for survival. According to CDC, you should keep one gallon of water per person. The water supply could be stored for a minimum of two days to a maximum of three weeks.

The amount of water you should have depends upon your budget, water needs, and family members. You can compute water by the general rule of every family member. Consider your water-dependent needs and habits, then add that amount for storage.

Conduct an Emergency Water Drill

There's one way you can determine how much water your family requires to stockpile for emergencies. Do a water drill.

An emergency water drill is a process where you go without running water for a specified period. During that time, you only consume your stockpiled water. At the end of the drill, you compute how much water you went through.

Water Conservation

Simply put, the better you are at saving water, the less water you'll need to stock for the entire family. Please don't wait until a disaster hits to learn different water conservation methods, or else you'll end up going through it quicker than you expected.

Here are some practical tips you can save water during emergencies:

- Use hand sanitizer while washing hands
- Use wet wipes to wipe yourself down instead of showering
- Use buckets to preserve Graywater when washing your hands, dishes, or clothes
- Make a tippy tap for washing hands with less water rather than pouring water straight over your hands.

Centers for Disease Control and Prevention. (2021). Creating and Storing an Emergency Water Supply. Centers for Disease Control and Prevention. Retrieved from https://www.cdc.gov/healthywater/emergency/creating-storing-emergency-water-supply.html

Pet Water Needs

Make sure you also don't forget to compute additional emergency water for your precious pets. To compute the quantity of water your dog or cat needs, take their weight in pounds and divide it by eight.

That's the amount of water they need every day in cups.

Long-Term Water Storage Solutions

So, you wanted to begin creating your emergency water supply. In that case, you will need a safe container to store it. The guideline is to employ food-grade plastic bottles. You can also use glass bottles so long as they have not stocked non-food items.

Stainless steel is a good option too, but you won't be able to treat your stored water with chlorine because it rusts steel.

No matter where you store the water, ensure you can properly and securely seal it. During an emergency, the last thing you want is bacteria or other contamination mucking up your stored drinking water.

Chapter 4:
STOCK FOOD SUPPLIES

Emergency food is a food supply stored in case of emergency. Disasters happen without warning, and you and your entire family can be cut from any food source. Hence, it's crucial to keep emergency food stored at your residence.

Plan Ahead

Having enough food supply in stock does not imply you should sit back and relax. Plan ahead of time and do not get caught unaware. It will help if you save money slowly just if your food gets destroyed by the disaster or goes bad.

Consider the Size of Your Family

You do not like to run out of supply early or skip meals to adapt to the accessible food. Do your calculations well and factor in your family size concerning daily food consumption.

Do you have children in the family? You cannot miss to include foods for your little ones. Do you have pets? Ensure they get well-nourished as well, even during low food supply.

Shelf Life

Various foods will differ in their shelf life. Some foods can be eaten even after their expiry dates, but that should not be the reason to overlook their shelf life.

Remember that the nutritional content for some foods will decline after their expiry date. That implies your body will receive lesser calories. Worse, you may need to eat more to keep your entire body nourished. High consumption will slowly impact your food projection food. Our point here is to get food supplies with longer shelf life.

Storage Space

It does not make sense to invest in too much food only to realize you don't have enough space to keep it. Ensure your storage could accommodate your food without crushing your place.

To prevent food spoilage, keep your supplies out of direct sunlight. Keep the humidity and temperature stable and low if possible. Some insects and moths will breed in wheat products, so always keep them sealed tight.

Ensure you rotate your supplies regularly and make sure you consume any food going out of date.

Dealing with Power Outage

What should you do when your power goes out? It's practical to consume first your perishable foods. If your freezer seal is airtight, you should get approximately twelve to twenty-four hours before the food spoils. Still, it will depend on the weather temperature outside.

What about heating and cooking food? In that case, you may need to resort to stove outdoor or camp oven.

Chapter 5:
PREP YOUR HOME

Natural disasters become more and more typical, making it more critical than ever for homeowners to ensure their home is protected in the event of an emergency. A few basic steps could make a big difference in safeguarding your family and home.

Check Your Home

Go around your house and make sure you know what everything does. It's quite shocking how most homeowners don't have any clue about the electrical aspects of their homes.

Get Homeowner's Insurance

Ensure you have homeowner's insurance which will cover the number of repairs you need in the event of a natural calamity. Basic policies may not cover specific damage caused by natural calamities, like a flood. It will help if you review the policy first and consider getting more coverage to protect your precious home.

Collect Insurance Documents

Your home insurance documents and other important papers must be part of your emergency kit and readily accessible in the event of a calamity. You need these documents for assistance and claims. Learn what your policy will cover.

Unplug All Electronic Appliances During Power Outage

Unplugging all your electronic appliances during a power outage will avoid any damage from the power surge after the power is restored. You may not know it yet, but you can save hundred dollars by just unplugging the appliances you don't use often.[4]

Prepare Your Garage Doors, Windows, and Doors

Do you live in a region prone to hurricanes? Then it is best to get your home ready for those scenarios with shutters. Brace all your doors and secure the garage door, which can blow away in a strong storm.

Bolt Your House to the Foundation

Ensure you have your home bolted to its foundation. Tornadoes, tsunamis, and earthquakes are some of the worst things that could take place in your home.

However, each of them has the likelihood to separate your house from its foundations. Hence, take precautions and hire an expert to bolt your home to its foundation.

Have a Contingency Plan

Are you the head of the household? You must take your responsibilities very seriously. Are you living in a high-risk area like a coastal town that's often in the path of hurricanes? Create a contingency plan with each of your families.

Not only should your family members know what to do during such emergencies. There must also be a protocol for responding when a calamity strike. That will offer everybody with valued peace of mind.

Family Handyman. (n.d.) Energy Vampires. Family Handyman. Retrieved from https://www.familyhandyman.com/project/energy-vampires/

Chapter 6:
STAY SAFE

Do you or someone you care for have health issues? Natural disasters could bring up a whole set of challenges. It is crucial to plan so you can be prepared for anything.

Nothing is more frightening than when a calamity strikes. Moreover, it is challenging to remember everything you need. That's especially true if that involves health-related information and supplies for someone you care for or yourself.

Determine Hazards in Your Home

Hazardous or dangerous objects and conditions around your house must be resolved to lessen the effect of a calamity. For instance, any leaky gas connections or frayed electrical wiring should be remedied.

Shelves should be thoroughly fastened, and oily solvents or rags need to be stored in metal containers properly.

Get First-Aid Supplies and Learn CPR Skills

All family members, including older kids, must learn basic first-aid skills. Keep in mind that prompt and efficient first-aid attention to medical emergencies like burns, heart attack, electrocution, bleeding, spinal injury, or choking could lower injuries and save lives.

Subscribe for Emergency Alerts

Stay updated with the information you can rely on by taking the time to subscribe for emergency alerts from your local government or other trusted sources. Unluckily, there could be all types of rumors and misinformation spread on social media and word of mouth when a calamity strike.

Safeguard yourself and your family by getting prompt facts straight from the source before, during, and after a disaster.

Understand How to Evacuate and Where to Go

Other calamities will need evacuation from your home, so you must be ready for that possibility. Determine if your home is in a high-level evacuation area. That could imply you need to evacuate in case of a wildfire, flooding, or hurricane.

Determine two or three choices for where you could go in case of emergencies, like a motel or a relative or friend's home.

Take Shelter

In the event of being requested to remain indoors, you can hide under a strong table. That will protect you from any falling debris.

Are you trapped in a hotel room? You can use duvets and pillows to your advantage and make a mound to hide under. The fabrics should cushion the impact of anything falling.

Chapter 7:
STAY CALM DURING POWER OUTAGE

In the middle of February 2021, a massive winter storm distressed Southern America, leaving millions of people without power or even safe drinking water.[5]

Are you ready to go through an emergency like that? It's frightening and somewhat overwhelming to imagine what may happen if the lights went out for a more extended period. But there are many things you can do to prepare for the worst.

Invest in a Solar Oven

You cannot light your gas oven without electricity. You can then invest in a solar oven that harnesses the power of the sun's UV rays to cook food just like a slow cooker. This device is an excellent addition to your emergency plan.

Personal Care & Bathing

Bathing will be another concern if the water stops flowing. Even if the water works, the water heater might not. That suggests a conventional shower or bath would be cold, a severe safety threat when the power is out.

Even if you don't need to shower regularly, you do need to keep yourself clean. That's why you need to store the proper supplies such as hand sanitizer, baby wipes, feminine hygiene products, diapers, and dry shampoo. You can also use water, soap, and a clean washrag for bathing.

---

[5] Booker, B., Romo, V. (2021). Winter Storm Leaves Many In Texas Without Power And Water. NPR. Retrieved from https://www.npr.org/2021/02/17/968665266/millions-still-without-power-as-winter-storm-wallops-texas

Lighting

You need to figure out how you will illuminate your home when the electricity goes out. Candles are a cheap option but can be risky, particularly if you have kids or pets in the house. Better lighting choices include:

- Headlamps
- Flashlights
- Solar or LED-powered lanterns

Communication and Technology

Damaged lines or storms could cause a power surge. It usually happens when the flow of electricity is interrupted and starts again. Surges typically make the overhead light flickers, but they can also destroy sensitive devices plugged into the walls, such as computers or mobile phones.

You can invest in surge protectors to safeguard your devices from any power surges that happen before an outage.

You can get a solar-powered radio to receive news, given that local radio stations can broadcast. It will also help if you get a solar charger to charge small electronic devices like your tablets and mobile phones.

Chapter 8:
EVACUATE

As with any emergency, dealing with a home evacuation with confidence comes down to being ready as possible. With the proper preparations, everybody could get to safety without mistakes along the way or getting too anxious.

Before you pick up and go, follow these tips. They are intended to safeguard you, your family and your pets and help avoid property damage.

Determine How to Leave the Area

It should begin by understanding the difference between emergency types and what to bring along.

For instance, in a house fire, everybody must leave the house as fast as possible without stopping to get anything. However, if there's some warning, like before a hurricane strikes, it may be best to get the emergency kit and go bags before exiting your home.

Equip Your Vehicle

It will also help if you determine the preferred evacuation vehicle in case the scenario requires leaving town. Your vehicle must be well-maintained, have about ¾ tanks of gas, and has an emergency kit in the trunk.

Secure Important Documents

Calamities have the potential to ruin important documents required to travel out of the country, drive a car, or prove who you are. To prevent such concerns, make copies and keep the originals in a safe location.

Find a Safe Place to Stay

City officials may open emergency shelters to residents during widescale evacuations. However, you cannot count on that alone. Hence, seek other alternatives that would work for the whole family.

If you have pets in the house, that's even more critical as most emergency shelters don't allow them to stay.

Remain Calm

Try your best to remain calm in this situation and find the best way of evacuation. You should learn where every family member is in the home so you can help them evacuate in the safest way possible.

On your way out, don't try to save any of your personal belongings. Doing that will raise your risk of getting injured.

Grab Your Fire-Starting Kit

Your fire-starting materials must be kept in a waterproof, sealed container. Materials include a lighter, matches, and some fire-starter. Various types are accessible online, but you can also create your fire-starter by cutting a candle in half. Extract the wick and wrap it in wax paper. Tuck those fire starters in the woodpile and light the wax paper with a match.

Chapter 9:
PROTECT YOUR VEHICLE

Are you living in a hurricane-prone area? You may understand how to prep your home ahead of any emergencies. But do you know that emergency preparedness must also extend to safeguarding your vehicles?

Having an inclusive safety plan is always better rather than being caught off-guard.

Take Photos

For personal and insurance purposes, take proof of your vehicle's condition before a calamity strikes. You may like to consider taking photos of the car's exterior and interior as you prepare before the storm hits.

Store Crucial Items

Keep copies of your vehicle's registration and insurance documentation in a safe place. Make extra copies of that and your fob or car key and distribute the keys to licensed drivers in the family.

Fill the Tank

Before a disaster hits, make sure you fuel up your vehicle. A crucial part of emergency preparedness is to get assistance when the situation subsides. A full tank of gas will help you reach your destination without having to make a stop for fuel.

Park Safely

Safeguarding your vehicle from a disaster involves protecting it from high water and winds. Park your car in your garage if possible. If you don't have one, try to park it close to a building. Stay away from parking under power lines or trees that could be blown down.

Always Check Your Vehicle

Safeguarding your vehicle from a disaster involves protecting it from high water and winds. Park your car in your garage if possible. If you don't have one, try to park it close to a building. Stay away from parking under power lines or trees that could be blown down.

Chapter 10:
DON'T FORGET YOUR PETS

A hurricane rushes through your state. A tornado strikes your city. A fire ruins your home. You have made it through safely, but what about your pets?

A pet should never be left behind to fend for themselves. If it's not safe for you, then it's also not safe for your pets.

Keep in mind that animals left during emergencies and natural calamities can become injured, starve, fall ill, drown from flooding, or die. They have less ability than people to escape threatening conditions, particularly if they're left in cages, tied up, or in a closed-up house.

Consider a Pet-Friendly Place to Stay

Look ahead for pet-friendly boarding facilities or hotels or make a housing exchange deal with a relative or friend.

Microchip Your Pets

Did you know that microchip identification is an excellent way to guarantee you and your pet are reunited in case of separation? Make sure you keep the microchip registration updated and add at least one emergency contact.

Keep a Tag or Collar on Every Dog and Cat

Keep different updated phone numbers on your pet's identification tag. Identification of those indoor-only cats is essential. They could easily escape if your home is damaged during a catastrophe.

Organize an Emergency Kit for Every Pet

Stock up on the things you may need during a disaster today, so you don't get caught impromptu. Here are the essential items you need to add to your pet's emergency preparedness kit:

- A week supply of freshwater
- A week supply of food
- Pet first aid kit
- Photos for identification
- Vaccination records
- Any medication
- Temporary ID tags
- Leash or carrier for every pet

Find Emergency Vet Facilities Outside Your Area

Emergency vet facilities may be closed if a calamity has impacted your community. Your pet may become ill or injured during the disaster. Ensure you are aware of how to access other emergency facilities.

Learn Where to Find Lost Animals

Animals often end up at a local shelter if they become lost during an emergency. Keep handy the phone numbers and locations of the animal shelters within your area.

Arrange for Temporary Confinement

Barns, fences, walls, and other physical structures may be ruined during a calamity. Have a plan for keeping your pet safely confined. You may need a kennel, tie out or crate.

Sometimes, when pets are evacuated to unknown locations, their fear and stress could result in illness or injury. Try to research more to ensure your pet's safety throughout an emergency evacuation.

Comfort Them

Your pet will value your calm presence and comforting voice following a calamity or while evacuated. You may also find it comforting to spend time with your pets.

Some pets, particularly cats, might be too frightened to be comforted. Engage with them on their terms. Other pets may find toys comforting.

Conclusion

Just imagine all the calamities that could strike your family, and home is not the safest place in the world. However, it is vital to be proactive and plan in case of an emergency strike.

Whether it's a hurricane, tornado, earthquake, flood, fire, tsunami, storm, or a global pandemic, emergency preparedness is the key to keeping you and your family healthy and safe.

We understand it will take some time to check off the items provided on this list. However, we know from our experience that you will feel better and more confident as you take every preventive action.

We cannot control emergencies and natural disasters. However, you can still empower yourself to do what you can to protect yourself, your family, your home, your car, your pets, and other important belongings.

You can still educate yourself and your family on safety and preparedness. One key is to create an emergency preparedness kit. Another way is to understand more about the various natural disasters that may strike in your area.

After you make a contingency plan, you can tinker with it a bit to address every disaster that could happen.

For more information on prepping for natural emergencies, you can visit the links below:

- **Ready.gov**

Check this website and double-check if your emergency supplies kit is complete.

- **Red Cross, Common Natural Disasters Across America**

Learn what types of disasters are more typical within your area. For instance, the Mid-Atlantic area is susceptible to hurricanes and winter storms. The West is at greater risk of tsunamis, volcanoes, landslides, earthquakes, and wildfires.

No matter where you live right now, prepare for extreme heat, floods, thunderstorms, power outages, and other types of natural disasters.

- **Red Cross**

Learn how to make an emergency plan in three simple steps along with free templates. You can integrate situations like things to do when your family members get separated, and you should evacuate your home.

To sum up, emergencies and natural disasters are stressful. Lucky for you, planning goes a very long way. Running drills and having an emergency kit ready to go means you are two steps closer to surviving a catastrophe intact.

This guide helped you prepare your home and family for disasters and emergencies to enable you to live comfortably and with peace of mind. It's always a great idea to put together an emergency plan and to have enough supplies on hand to keep you and your family safe.

So, what are you waiting for? Do not waste any critical moments panicking. Plan today—it's better to be safe than sorry.

We hope you find this guide highly informative and fun to read at the same time. You now have all the tools and information you need to be prepared for any emergency. Go ahead and use them efficiently!